This is the earth.

This is a globe.

A globe is a model of the earth.
It is round like the earth.
We use a globe to find places.

A _____ is a round model of the earth.

Color the land green.
Color the water blue.

Note: After doing this page with your students, have them explore a real globe. You will find more globe activities on the inside back cover of this unit.

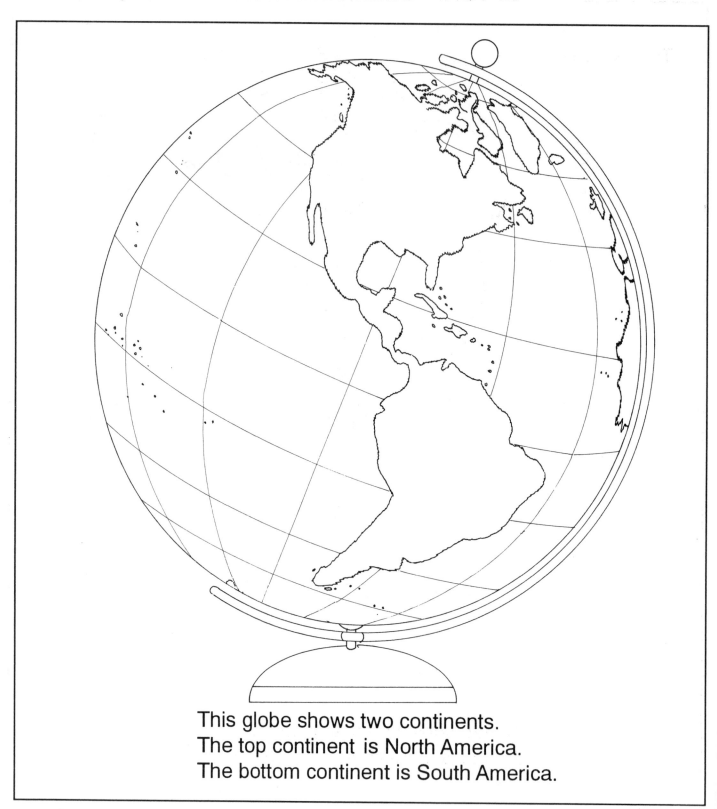

This globe shows two continents.
The top continent is North America.
The bottom continent is South America.

Color the continents.

North America - orange

South America - yellow

If you live in North America make a red X on it.

If you live in South America make a black X on it.

   Beginning Geography

This is a map.

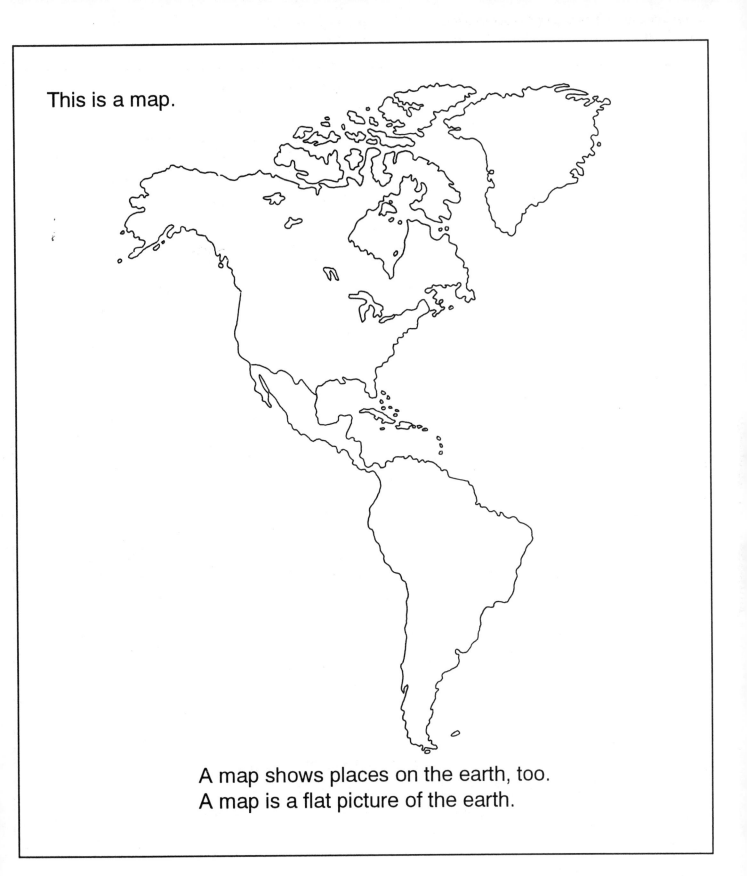

A map shows places on the earth, too.
A map is a flat picture of the earth.

A _____ is a flat way to show the earth.

Color the land green.
Color the water blue.

                       Beginning Geography

paste

paste

paste

North, south, east, and west are directions.
These directions help us find places on maps.

paste

east	west
north	south

Note: Have a real compass in class. Help children to locate and label north, south, east, and west in your classroom.

This is a compass.
You use it to find directions.
A compass always points north.

This is a compass rose.
You use it to find directions on a map.
A compass rose points north, south, east, and west.

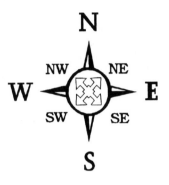

Which compass goes here?

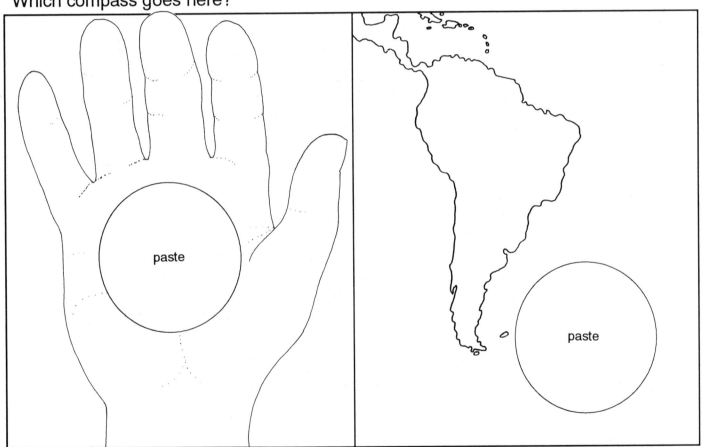

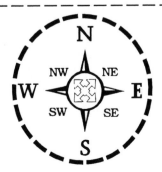

5 Beginning Geography

Using Maps

Around the Neighborhood - Reproduce the map on page 7.

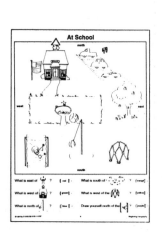

Give oral directions to children guiding them from place to place on the map.

1. Practice by having them use their finger to move short distances. Give directions such as these.

    "Start at the school. Go to the playground."
    "Start at the pond. Go to the vacant lot."

Continue with this activity until your students feel comfortable with the map.

2. Review the directions north, south, east, and west.

    "Put your finger on the playground. Move your finger north."
    (Practice moving south, east, and west.)

    "Put your finger on the park. Go south two blocks. Turn east.
    Where are you now?"

Continue with this activity until your students feel comfortable moving in different directions.

3. Have your students take a pencil and draw a line along the streets as you give oral directions such as these.

    "Start at the house in the corner."
    "Draw a line to the car wash."
    "Go on to the pool."
    "Now go to City Hall."
    "Go on to the school."
    "Now go to the library."
    "Stop at the park."

At My School - Reproduce the map on page 9.

Have children look at the map as you ask questions such as these.

    "Is the flag on the west or on the east side of the school?"
    "Is the drinking fountain north or south of the tether ball?"
    "Will you go east or west to get from the bars to the sandbox?"

You may repeat these activities many times using the same reproducible maps and giving different directions.

Note: Have children locate and color places on the map.

Around the Neighborhood

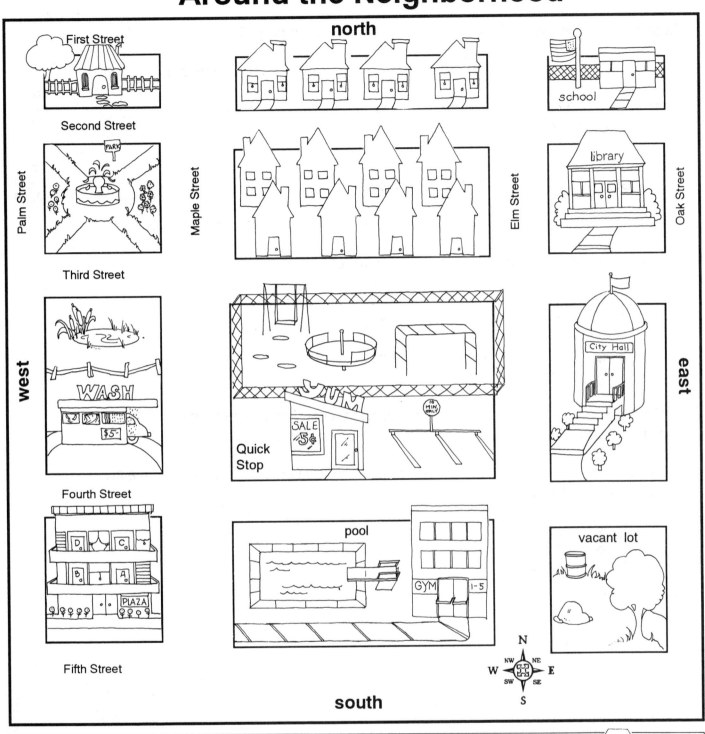

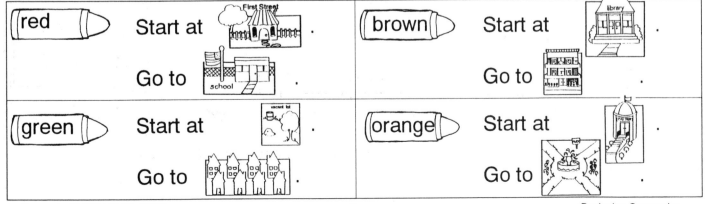

7

Beginning Geography

Symbols

These are symbols.
Symbols stand for real things.
Here are some symbols you may see on a map.
What do you think they stand for?

marina

airport

campfire

food

school

homes

restrooms

church

information

railroad

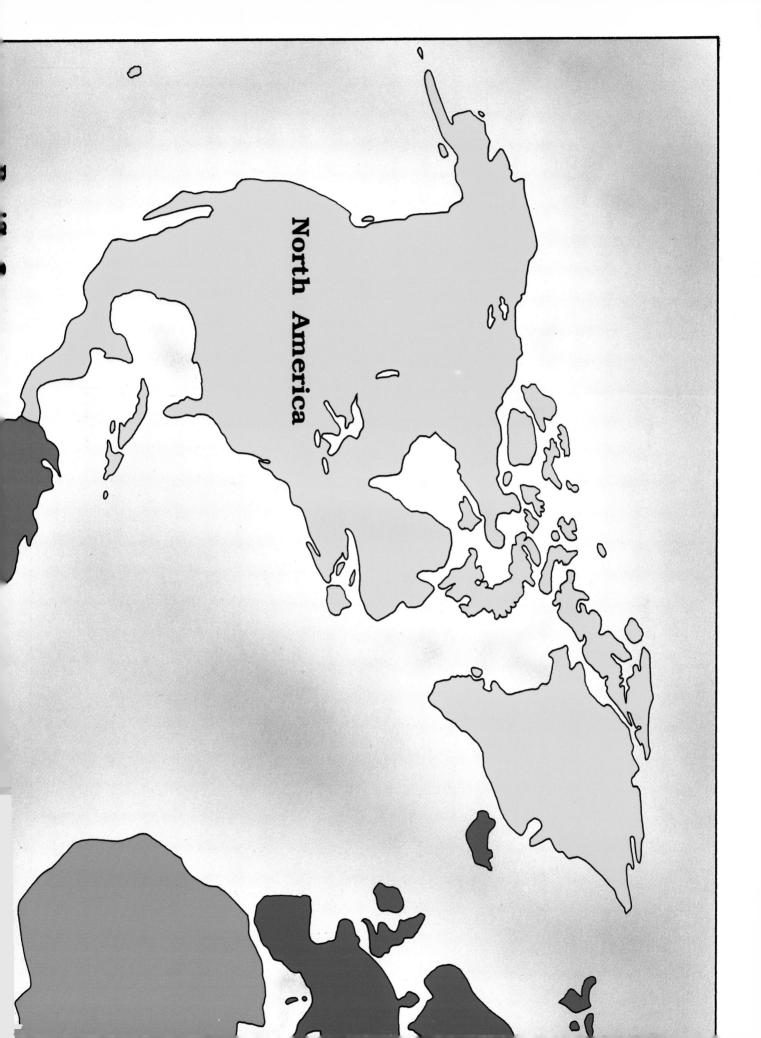

North America

A Map of

Fairy Tale Land

Arctic Ocean

Asia

Europe

Africa

Australia

N
W → E
S

Antarctica

EMC 259 Beginning Geography

outh

A World Map

South America

Atlantic Ocean

Antar

Note: After using this page you may want to have your students work together to make a big map of their own school.

At School

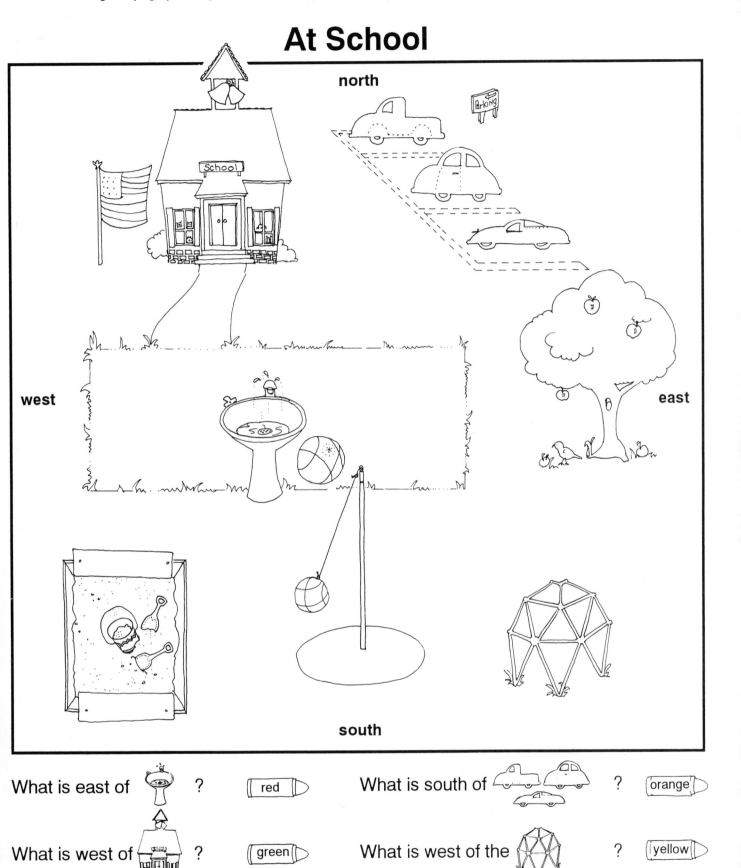

north

School

Parking

west

east

south

What is east of 🖉 ? [red]

What is west of 🏫 ? [green]

What is north of ⚽ ? [blue]

What is south of 🚗 ? [orange]

What is west of the 🔵 ? [yellow]

Draw yourself north of the 🪣 [purple]

In the Classroom

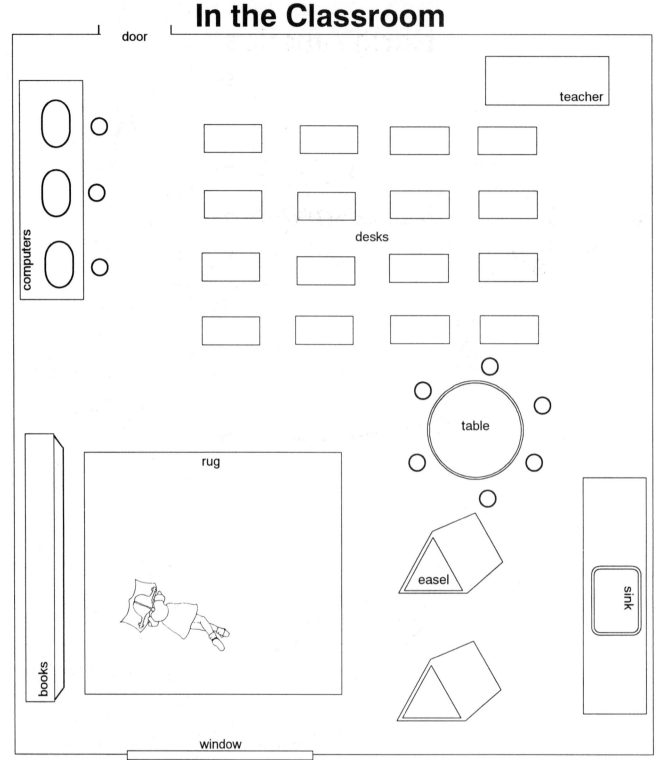

Follow directions:

Make a red apple on the teacher's desk.

Draw a red line from the teacher's desk to the rug.

Make a green box on the bookcase.

Color blue water in the sink.

Write your name on one of the desks.

Color your favorite color on an easel.

Draw a yellow line to show the shortest distance between the window and the door.

Draw a purple line to show a long route to get to the door from the window.

North America

This is a map of North America.
It is divided into countries.

Color the three biggest countries.

Canada - green
United States of America - purple
Mexico - orange

In the United States

This map shows the United States of America.
It is divided into states.

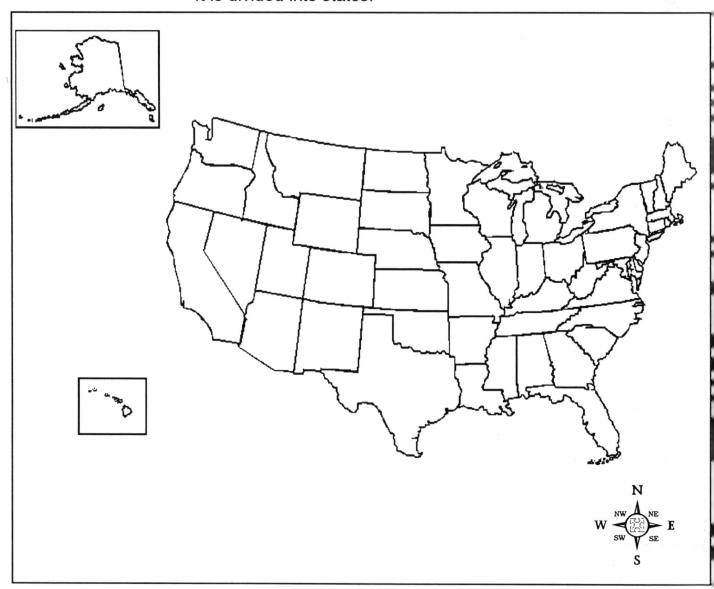

1. Color California red.

2. Put a brown X on Texas.

3. Make a green circle around Florida.

4. Put an orange X on Hawaii.

5. Color Alaska blue.

Help Find the Buried Treasure

Follow the map from the ship to the buried treasure.

Start at the ship.
Go north until you get to the lake.
Go west until you reach the cave.
Sh! Don't wake the bear inside.
Go south until you get to the river.
Walk across the log. Don't fall in!
Go east. Walk behind the volcano.
Turn and walk south until you come to the village.
Go north until you get to the three rocks.
Look under the dark rock.
Hurray! You have found the treasure.

Beginning Geography

Note: Reproduce this map of the continents for your students. Begin by having them color the continents to match the poster of the world. Repeat the directions you gave on page 16 for finding places on the poster of the continents.

The Continents of the World

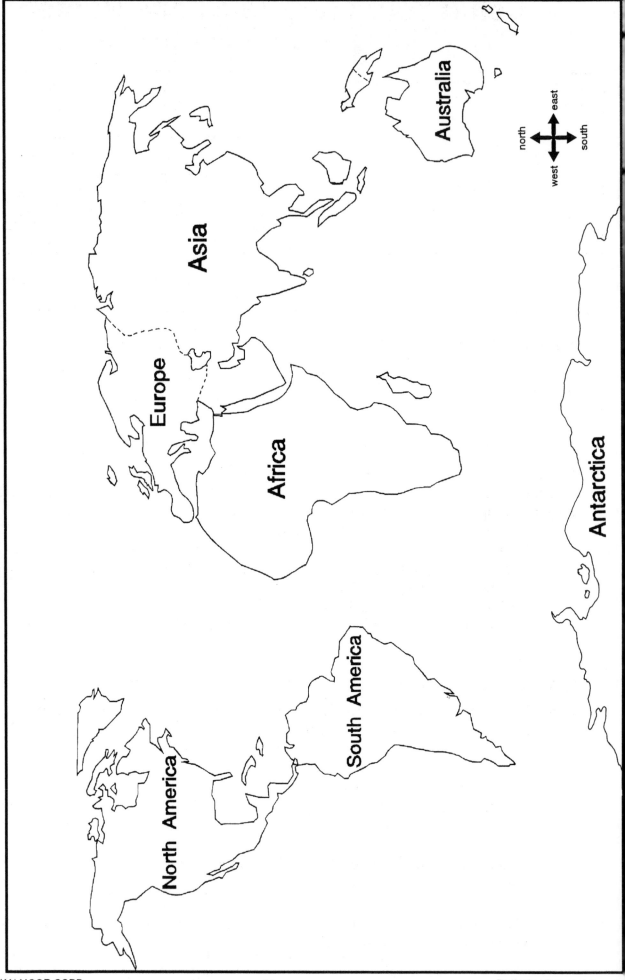

14

Beginning Geography

Using Your Fairy Tale Land Poster

Use the map of fairy tale land to practice map skills.

1. Naming Places on a Map

Have your children identify and locate the fairy tale characters and their homes on the map.

2. Naming Directions on a Map

Review where you find north, south, east, and west on a map. You may make this more challenging by covering up the words printed on the map and having your students use only the compass rose to figure the directions. Ask questions such as these.

> "Does Grandma live north, south, east, or west of the Little Red Hen?"
> "Do the Three Bears live north or south of Jack and his mother?"
> "In which direction is the fox swimming across the river?"
> "Is the Big Billy Goat Gruff on the west or east end of the bridge?"
> "Is the Little Red Hen's windmill on the south or the north side of her yard?"

3. Moving Around a Map

Have children follow directional words to move from one place on the map to another. You might give directions such as these.

> "Start at the Little Red Hen's gate. Follow the road turning in these directions - go east, now south, then east again to the first house, now north. Are you at the Little Pig's straw house?"

> "Start at Red Riding Hood's house. Follow the road turning in these directions - west, south, then west and into Grandma's house."

> "Tell me how to get from the Little Pig's brick house to Jack's house. Use the words north, south, east, and west to direct me."

Using Your Poster of the World Map

1. Vocabulary Development

Help your students to name and locate continents
and countries on this world map.

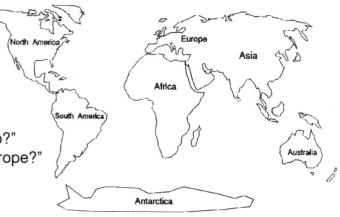

"How many continents can you count?"
"Which continent is ___(color word)___?"
"Which continent is at the bottom of the map?"
"Is North America bigger or smaller than Europe?"
"On which continent do you live?"

2. Naming Directions on a Map

Review where you find north, south, east, and west on a map. You may make this more
challenging by covering up the words printed on the map and having your students use
only the compass rose to figure the directions. Children may be able to point to the
places on the map before than can give you the names. Ask questions such as these.

"Which continent is west of Asia?"
"Which continent is south of North America?"
"Is Europe east or west of North America?"

3. Moving Around a Map

Have children follow directional words to move from one place on the map to another.
The age and ability of your students will determine how easy or difficult you need to
make the questions.
You might give directions such as these:
Easy:
"Start at the red continent. Go north. What is the color of the first continent you
reach?" "Do you know its name?"
"Go from the green continent to the yellow continent. Which direction did you
travel?"

Difficult:
"Start at Australia. Go north until you reach a continent. Name the continent.
Now turn east. Travel east until you reach a new continent. What is the name of
that continent?"

4. Reproduce the map on page 14 for each child.

Have them color the continents to match the poster. Repeat any part of activities one
through three, having the children follow along on their own maps.